ESSAYS ON MORAL REALISM